The Introduction of Bilalian Ryu Jujitsu
By Dr. Abdul Hakim Bilal, Founder

ntroduction

d like to give thanks to The Most High and to my Bilalian Ryu Jujitsu family. Many thanks to
l of my great Masters whom have taught me. Many thanks to my uncle, Grand Master
illie McPherson, and all the McPherson Family, my brother Angelo Hickman, Master Clyde
aint" Bowal, Master William Lawton, Master Israel Gonzalez, my friends, who know me, you
ow who you are and all of the Martial Artist around the world who have inspired me.

hank you.
rand Master Blackhawk Sancarlos,Grand Master Papasan Canty
laster Alan Goldberg, Grand Master Billy Davis
rof. Amin Hassan, Sensei Reginald Brown ,Grand Master Allah Freedom,
rand Master George Crayton, Grand Master Reno Morales,
rand Master Rico Guy, Grand Master Michael Willett,
ushinda Lamarr Thornton, Prof. Jonathan Stewart,
appasan Jack Stern, Master Mo Mahaliel Bethea,
laster Earl Woodbury, Master John Gray, Sensei Yoko Ushioda Gary,
rand Master Pop Edwards Sr. Jr. , Shihan Darlene Defour,
laster Lil Katan Jones, Shihan John Wheaton, Shihan Candy Soto,
r. Sharif Wali, Grand Master Abdul Musawwir,
r. Rocky Farley, Hanshi Rich Diaz ,Grand Master Hassan Kaleak,
laster Dino Blanche, Master Abdul Aziz, Master Lee Ireland,
oke Dexter Cockburn, Master Stacey Johnson, Dr. Joe Parrish,
r. Cliff L. Thomas, Master Terry D. Richardson, Grand Master Bahiy
luhammad, Master Tim Elder ,Master Carlos Ampudia Sr. Jr. Jessica,
ensei Maria Ham, Grand Master Klaus Schuhmacher,Dr. Carlos Diaz ,
rand Master John W. Smith, Prof. Abdulmuhsiy Abdurrahman,
rand Master Lil. John Davis, Robert Handley Hanshi

BIOGRAPHY OF SOKE
ABDUL HAKIM BILAL

I was born on July 30, 1947 in Harlem New York City, USA.
MY uncle , Willie McPherson was the first man ever to introduce me to the martial arts at the age of 9. He would show me locks, sweeps and chokes, an art known as jujitsu 1956.
I began My training Goju Ryu in North Carolina 1965.
After leaving North Carolina I moved to New York City
I met a man named Master Zeus in Harlem and began to train with him in Goju Ryu . His teachers name was GrandMaster Yamaguchi.
I studied with him for several years 1969 I met many martial artist, one being Master Thomas Bodie, who taught me more about Goju . He was a great mentor to me . Another brother I met, known as Monster man Eddie also took me under his wing in Goju as well.
My training began with Professor Moses Powell I studied with him for several years. I also studied Kung Fu Woo Su .also studied Fuji Jujitsu.
Under Master Raheem Alemeen and Master Neil O Farrell.

 MY ACHIEVEMENTS INCLUDE NUMEROUS INDUCTIONS IN HALL OF FAMES ALL AROUND THE WORLD
I am certified in the Unified Sokeship Association Founder of Bilalian Ryu Jujitsu . A member of Eastern U.S.A. int ,l Martial Arts Association. World Karate Union Federation, a Board member of the Worldwide Martial Art Hall of Fame. I am recognized by Professor Ed Brown,s Martial Art Masters, Pioneers / Legends Hall of Fame . United States Martial Arts Association Hall of Fame .Action Magazine Hall of Fame .World Head of Society, Woma Sokeship ,Korean Yodo and Hapkido Association, GO SEI GOJU, Doctor of Philosophy in Asian Martial Arts, World Martial Arts League, in
Asia,Europe,U.S.A.,Latin,America,Africa,Brazil,Carbbean,Australia. World Soke Head Council OF Japan.
U.S.A. Martial Arts Hall of Fame.
BUDO International Hall of Fame. Insider Martial Arts Magazine.
Hansu Kido Hae Hapkido Black Belt Certificate Presented to Abdul Bilal 6th Dan , International Sokeship Council The Rank of Hanshi,
International Founder and Headmaster Society.

Published by Lulu.com

780557 369287
ISBN 978-0-557-36928-7

90000

CHAPTER 1 - Wrist Techniques

Figure 1

Step in with right foot and right hand grabs opponent's right hand.

Figure

Step right leg back into horse stance position while applying wrist lock (quota gaiesh).

Figure 3

Step in with right foot and right hand grabs opponent's right hand, left leg steps down into horse stance position while applying wrist lock (quota gaeish).

Figure 4

Step in with left leg and left hand grabs opponent's right hand.

Figure 5

Apply wrist lock, dropping opponent to the ground.

Opponent grabs your right wrist with his left hand.

Figure 7

Take your left hand and grab opponent's left hand, thumb to thumb.

Figure 8

Keep opponent's left hand locked as you go outward pointing downward.

Figure 9

Opponent grabs both your hands from behind.

Figure 10

Step out with your right leg, bringing your right energy hand up and your left energy hand downward.

Figure 11

Step and pivot your left leg forward rotating your right energy hand downward and your left energy hand on opponent's right arm, pushing forward.

CHAPTER II

Choke Techniques

Figure 12

Opponent front chokes you, you strike 2 inches under opponents elbow.

Figure 13

Opponent responds to strike.

Figure 14

Opponent front chokes you, you break choke with pyramid block.

Figure 15

Apply one double strike to collar bone.

Figure 16

Grab opponents head and knee strike to head.

Figure 17

Follow up with kick to groins.

Figure 18

Opponent side chokes you, you raise your elbow up striking opponents chin.

Figure 19

Drop down into horse stance position, clawing face.

Figure 20

Reverse punch to opponent's solar plexus.

Figure 21

Opponent is mugging you from behind, grab both of his hands, stomp both opponents' feet.

Figure 22

Shift towards the outside of opponent's body, elbow strike to opponents solar plexus.

Figure 23

Strike to opponents groins

Figure 24

Grab opponent's leg, look over your shoulder, kick to groins.

CHAPTER IV - Energy Hand Blocks

Figure 25

Opponent round house strikes to your head, you step out same leg same hand with energy hand block.

Figure 26

Follow up with opposite hand with reverse punch to face.

Figure
27

Follow up with opposite hand with reverse punch to solar plexus.

CHAPTER V
Overhead Energy Hand Block and Grab

Figure 28

Side step to the outside of your opponent with a circular energy hand block, grabbing opponent's attack hand.

Figure 29

Follow up with a roundhouse kick to opponent's groins.

Figure 30

Followed by a hop in sweep, taking opponent down to ground.

Figure 31

Followed by a double punch to face.

CHAPTER VI - Chokes and Takedown Techniques

Figure 32

Opponent comes with an overhead strike to your head.

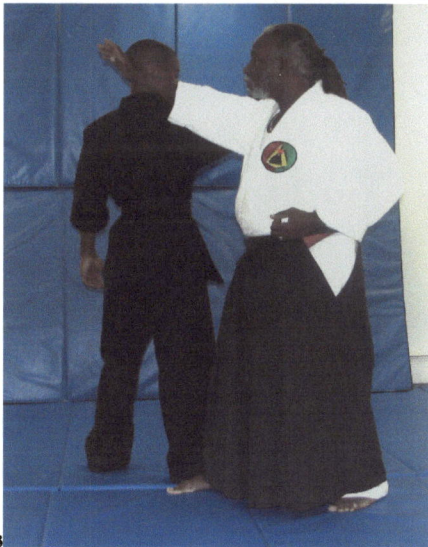

Figure 33

Step with your left leg pass opponent with your right energy hand upward.

Figure 34

Step around opponent, left fist to opponent's lower spine.

Figure 35

Push and pull, opponents head into your right shoulder.

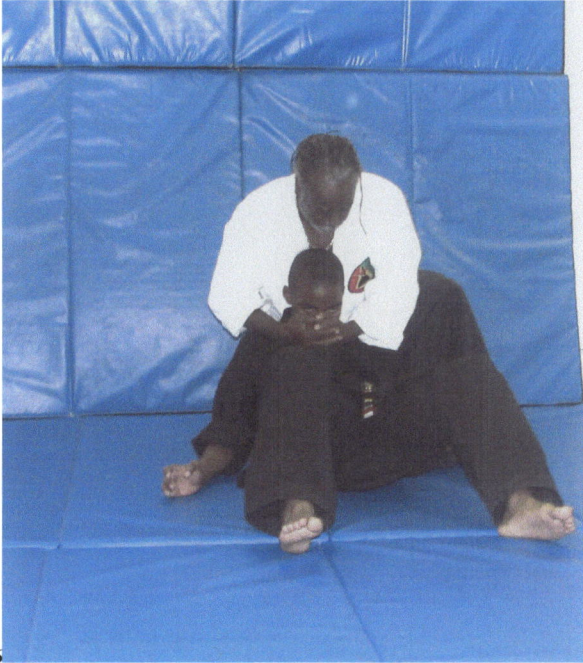

Figure 36

Choke and smother.

CHAPTER VII - Knife Techniques

Figure 37

Opponent stabs toward your solar plexus area.

Figure 38

Step back into horse stance position, circular hammer block, making opponent release weapon.

Figure 39

Thumb to eye.

Figure 40

Grab wrist (quota gaiesh)

Figure 41

Step around inside opponent's body.

Figure 42

Still applying wrist lock, take down.

Figure 43

Opponent round house slashes towards face. Step with your left leg pass opponent with your right energy hand upward inside block.

Figure 44

Follow up with opposite hand with reverse punch to face.

Figure 45

Left hand grabs opponent's knife hand.

Figure 46

Step with right leg inside opponent's body.

Figure 47

Slashing opponent across neck.

Figure 48

Sweep.

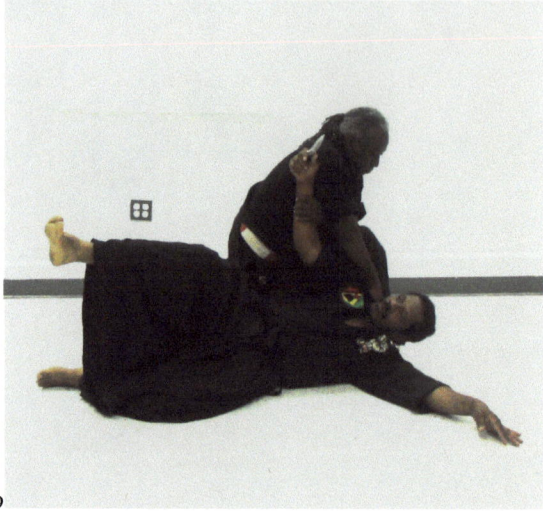

Figure 49

Upon take down, control opponents knife hand, punching to face.

Figure 50

Lock knife hand across opponent's body while dropping knee to head.

Figure 51

Opponent stabs with an overhead strike to your head.

Figure 52

Side step with your left leg, letting opponent pass you while your right hand grabs opponent's knife hand.

Figure 53

Right round house kick to opponent's groins or solar plexus.

Figure 54

Hop in, grab opponents shoulder and sweep.

Figure 55

Side step tai sabaki #4, letting opponent pass with knife hand, shuto back of opponent's neck. (Followed by figure 54)

Figure 56

Keep control of knife hand with both of your hands. Follow up with left heel choke across opponent's neck.

Figure 57

Opponent round house slashes towards face, step with your left leg with left circular energy hand block.

Figure 58

Grab opponents knife hand and pass to your other hand, keeping wrist lock (quota gaiesh).

Figure 59

While applying wrist lock (quota gaiesh), bring knife hand over head.

Figure 60

While in control of wrist, drop down on right knee, while bringing knife hand downward. Opponent is taken down.

Figure 61

Maintaining opponent's knife hand with wrist lock (quota gaiesh).

Figure 62

Pass wrist lock (quota gaiesh)

Figure 63

Breaking wrist causing opponent to turn over face down, taking the knife.

CHAPTER VIII – LAPEL GRAB TECHNIQUES

Figure 64

Opponent grabs your left lapel with his right hand

Figure 65

Grab opponent's right hand with your right hand.

Figure 66

Wrist lock opponent's right hand, keeping wrist lock pinned to chest, while your left hand slides inside of opponent's elbow.

Figure 67

Bring left elbow into you, while bringing your right knee up.

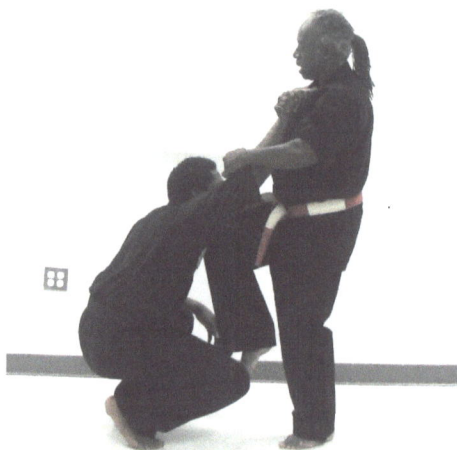

Figure 68

Right knee strikes opponent.

Figure 69

Two hands grab your lapel, as both your hands come up.

Figure 70

Your left hand holds opponent's right hand, stepping with your right foot, while striking opponent's face with a right inside elbow strike.

Figure 71

Opponent grabs your lapel with his right hand, wrist lock opponent's right hand with your right hand, while left energy hand bridge his right wrist.

Figure 72

Opponent grabs your lapel with his right hand, hold opponent's right hand with your right hand, while your right hand heel palm strikes to opponent's nose.

Figure 73

Opponent grabs your lapel with his right hand, left downward punch to opponent's solar plexus.

Figure 74

Opponent grabs your lapel with both of his hands, left downward punch to the inside of opponents left arm.

Figure 75

Opponent grabs your lapel with his right hand, left downward punch to opponent's left lower rib.

CHAPTER IX – EXERCISES

Up and down : Wrist, stomach, hand.

Arm movements: Up, side

Arm movements: Back, Front between legs.

Shift: Left to Right

Neck Rotation: Up-down, left-right, circular rotation left and then right.

Leg stretching: Walk it out. Hold flat, Hold on Heels.

Shift to your right –Hold.   **Shift to your left –Hold.**

Floor Exercises – On your back

V-Ups

Sit-Ups

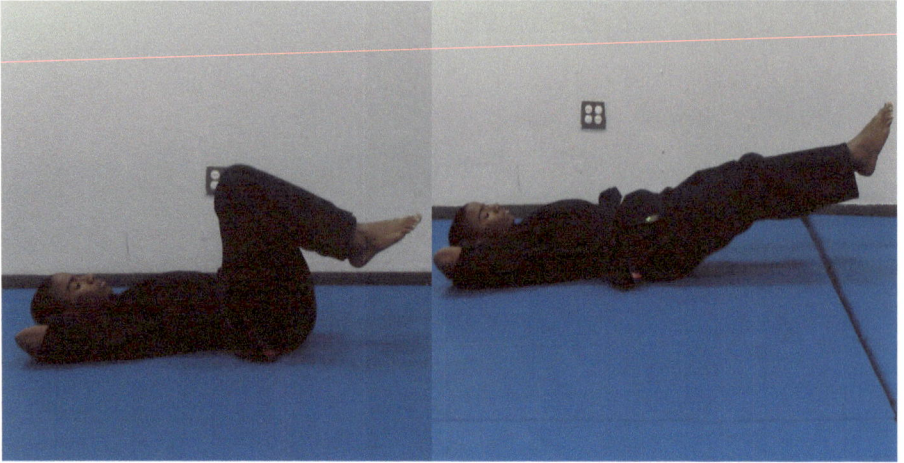

Knee-Ups

PUSH UPS

Flat Hand

Fingers

Knuckles

Pyramid

Wrists

Wrist Exercises

Quota Gaiesh

Chicken Neck

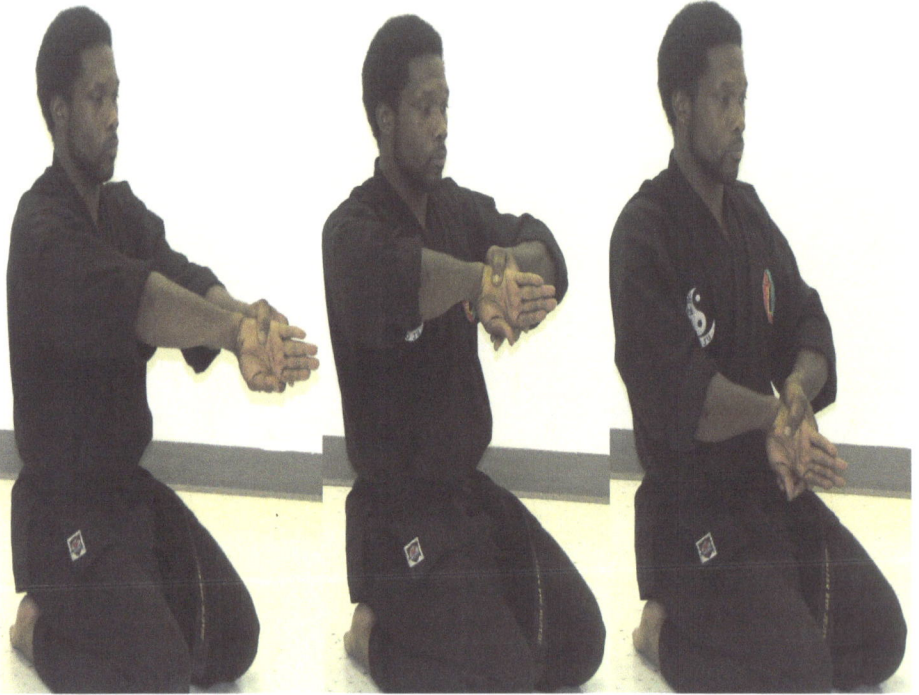

Thumb Down – Pull in, circular back out (repeat), changing hands.

Reach   Pull

CHAPTER X - Tai Sabaki – BASIC

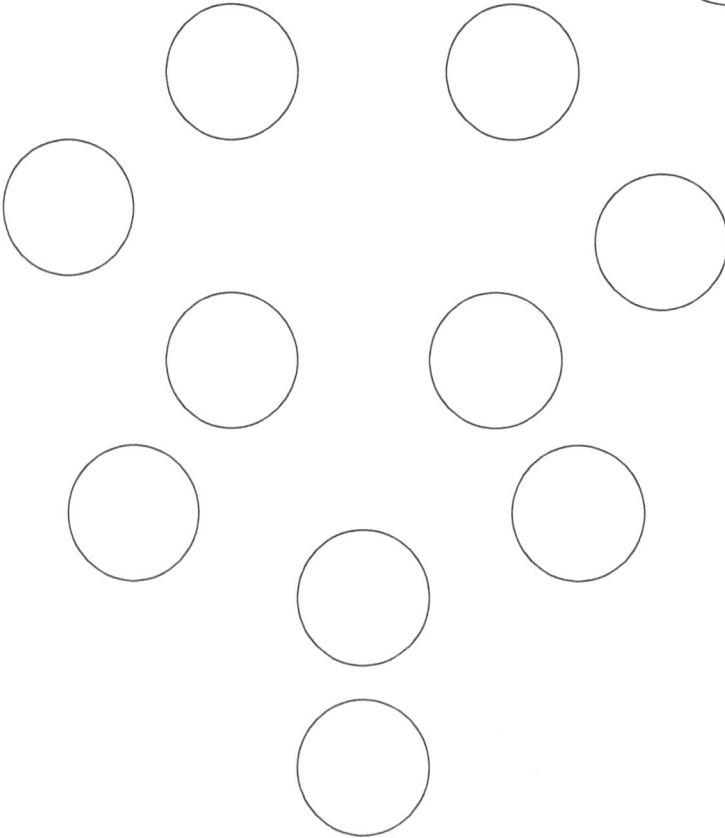

www.ingramcontent.com/pod-product-compliance
Lightning Source LLC
Chambersburg PA
CBHW041312210326

41599CB00003B/86